SCIENTIFIC AMERICAN | EDUCATIONAL PUBLISHING

SCIENTIFIC AMERICAN INVESTIGATES REPTILES

IGUANAS

BY JAGGER YOUSSEF

Published in 2026 by The Rosen Publishing Group
in association with Scientific American Educational Publishing
2544 Clinton Street, Buffalo NY 14224

Cataloging-in-Publication Data
Names: Youssef, Jagger.
Title: Iguanas / Jagger Youssef.
Description: Buffalo, New York : Scientific American Educational Publishing, an imprint of Rosen Publishing, 2026. | Series: Scientific American investigates reptiles | Includes glossary and index.
Identifiers: ISBN 9781725352919 (pbk.) | ISBN 9781725352926 (library bound) | ISBN 9781725352933 (ebook)
Subjects: LCSH: Iguanas–Juvenile literature.
Classification: LCC QL666.L25 Y68 2026 | DDC 597.95′42–dc23

Portions of this work were originally authored by Kathleen Connors and published as *Iguanas.* All new material in this edition is authored by Jagger Youssef.

Designer: Andrea Davison-Bartolotta
Editor: Therese Shea

Photo credits: Cover, p. 1 Artush/Shutterstock.com; p. 5 (top) I Wayan Sumatika/Shutterstock.com; p. 5 (bottom) Aspects and Angles/Shutterstock.com; p. 7 fotosteve/Shutterstock.com; pp. 9, 15 Milan Zygmunt/Shutterstock.com; p. 11 Cezary Wojtkowski/Shutterstock.com; p. 13 GUDKOV ANDREY/Shutterstock.com; p. 17 Ralfa Padantya/Shutterstock.com; p. 18 MDay Photography/Shutterstock.com; p. 19 (inset) Maridav/Shutterstock.com; p. 19 (main) travelview/Shutterstock.com; p. 21 (top) AndreySkat/Shutterstock.com; p. 21 (bottom) alexcursion/Shutterstock.com.

Printed in the United States of America

CPSIA compliance information: Batch #CSSA26. For Further Information contact Rosen Publishing at 1-800-237-9932.

CONTENTS

COLORFUL CREATURES 4

THE IGUANA BODY 6

WHERE TO FIND THEM. 8

HOT AND COLD 10

EATING (MOSTLY) GREEN 12

FINDING A MATE. 14

IGUANA HATCHLINGS 16

UNIQUE LIZARD 18

PROTECTING IGUANAS 20

GLOSSARY 22

FOR MORE INFORMATION 23

INDEX 24

Words in the glossary appear in **bold** type the first time they are used in the text.

COLORFUL CREATURES

Iguanas are lizards, a kind of reptile. Iguanas can be different sizes. Some species, or kinds, are fewer than 6 inches (15 cm) long, and some are larger than 7 feet (2.1 m) long. Males are often larger than females.

Iguanas can be many colors too. Many are green, brown, or yellow, but others are more gray or blue. Still others can change their look, becoming orange and red sometimes.

Still, all iguanas have certain features in common. Read on to learn about amazing iguanas!

Iguanas can be different colors and have different markings too.

FUN FACT

LIZARDS ARE A FAMILY OF REPTILES. REPTILES ARE ANIMALS WITH A BACKBONE THAT ARE COVERED IN SCALES OR BONY PLATES. THEY BREATHE AIR AND USUALLY LAY EGGS.

THE IGUANA BODY

Iguanas have **stout** bodies covered in scales. Different shapes and sizes of scales cover different body parts. Iguanas also have a **crest** down their back. The **spines** on it look a bit like a comb.

An iguana has a flap of skin under its chin called a dewlap. All iguanas have long tails that are about half of their total body length. Their four short legs end in five toes with sharp claws.

FUN FACT

IGUANAS HAVE LARGE, ROUND SCALES ON THEIR CHEEKS. THESE ARE CALLED SUBTYMPANIC SHIELDS.

Iguanas use their tongue to smell, just like another kind of reptile—snakes!

WHERE TO FIND THEM

Iguanas live mostly in the Western **Hemisphere**, especially Central and South America, Mexico, the Caribbean, and southwestern United States. Two species live on the Galápagos Islands in the Eastern Pacific Ocean, and three species are found in Fiji in the South Pacific. People have brought iguanas to other places as well.

Some species of iguanas live in trees. The green iguana can be found in hot, wet rainforests. The desert iguana likes the dry heat of deserts. Still other iguanas live on the ground in rocky or swampy **habitats**.

Green iguanas can survive falling 40 feet (12.2 m) from a tree branch! Living high in trees helps them avoid some predators.

FUN FACT

SCIENTISTS HAVE COUNTED ABOUT 35 SPECIES, OR KINDS, OF IGUANAS SO FAR.

HOT AND COLD

Iguanas are cold-blooded, just like other reptiles are. That means their body temperature depends on their **environment**. They spend a lot of time not moving in the sun. This is called basking. They need the sunlight to keep their inner body warm.

Temperature can affect an iguana's color. When they're cold, iguanas may become slightly darker. This helps them take in more heat from the sun. If the temperature is too warm, these lizards might become lighter. They may also seek shade if they become hot.

A black iguana basks in the sun in the Central American country of Belize.

FUN FACT

IGUANAS THAT GET TOO COLD STOP MOVING. THEY MAY FALL FROM TREES! WHEN THEY HEAT UP, THEIR BODY MOVEMENTS RETURN TO NORMAL.

EATING (MOSTLY) GREEN

Most iguanas are herbivores, or plant eaters. They like to munch on tree leaves, buds, flowers, and fruits. Iguanas that live in trees can easily find these things to snack on.

Iguanas that don't live in trees eat what they can find. Desert iguanas eat cactuses, bugs, and sometimes carrion, or dead animals. The marine iguana eats **algae** in the ocean. Young iguanas have been known to eat worms and even small birds.

FUN FACT

IGUANA PREDATORS INCLUDE HAWKS, OWLS, SNAKES—AND PEOPLE! SOME PEOPLE RAISE IGUANAS TO EAT THEIR MEAT.

This Galápagos land iguana eats a prickly pear cactus.

FINDING A MATE

When it's time for iguanas to find a **mate**, they may change their look and their behavior, or way of acting. Both males and females may turn an orange-red color, though males commonly become brighter than females. This can happen all over their body or just on parts, such as their dewlap, spines, or legs.

Males may also fight each other, bob their heads, or do a movement that looks like push-ups. They may move their tail in a certain way too. These things attract, or draw the attention of, a female mate.

Green iguanas may turn bright orange during mating season. This is usually toward the end of the rainy season where they live.

FUN FACT

IGUANAS OFTEN LIKE TO LIVE ALONE. THEY COME TOGETHER IN PLACES WITH A LOT OF FOOD OR A GOOD PLACE TO BASK IN THE SUN.

IGUANA HATCHLINGS

Female iguanas lay eggs after mating, at the beginning of the dry season. How many eggs depends on the species. Some lay as few as five and others as many as 50. These groups of eggs, called clutches, are buried in nests dug in the ground. The soil around them keeps the eggs warm.

After about two to three months, the iguana eggs hatch. Baby iguanas, called hatchlings, use their claws to dig out. They commonly look like little copies of their parents. However, they're usually a brighter color.

Iguanas with more predators tend to lay more eggs at a time. Eggs and hatchlings are common prey for some animals.

FUN FACT

IGUANA EGGS ARE LEFT ALONE BY THEIR PARENTS. HATCHLINGS HAVE TO TAKE CARE OF THEMSELVES TOO.

UNIQUE LIZARD

The marine iguana of the Galápagos Islands might be the most **unique** iguana species. It has a scary face and spiky scales like a movie monster, but it's harmless.

Marine iguanas may have a strange-looking white "wig." That's because they eat a lot of salt when they eat their favorite food in the water—algae. They can hold their breath for up to an hour underwater! Marine iguanas get rid of the extra salt they take into their bodies through their noses.

FUN FACT

THE MARINE IGUANA IS THE ONLY LIZARD THAT SWIMS IN THE OCEAN.

Marine iguanas can look like they're sneezing out salt! It dries on their head.

PROTECTING IGUANAS

Some iguana populations are healthy, but others are in danger of extinction, or dying out completely. The rock iguanas of the Caribbean are among the most **endangered** of all lizards. They began to die out as more people settled on Caribbean islands and destroyed their habitats for buildings and roads. People's pets also eat iguana eggs and hatchlings.

Zoos are helping keep some iguana species alive. However, these lizards mate much more easily in the wild. People need to make sure iguanas have wild habitats in the future!

More Iguana Fun Facts

They have excellent eyesight.

They have a body part on the back of their head called a third eye. It can sense changes in light and movement.

They live about 15 years in the wild.

They can lose their tail to escape predators, and it will grow back!

They use their tail for protection, balance, and swimming.

Some people keep iguanas as pets. They're not easy to keep happy and healthy though.

This road sign in Guadeloupe, in the Caribbean, says "Slow down, iguanas" in French.

GLOSSARY

algae: Living plantlike things that are mostly found in water.

crest: A raised part of an animal body, such as a ridge that runs along the back.

endangered: In danger of dying out.

environment: The air, water, and land around and on which a plant or animal lives.

habitat: The place where an animal or plant lives in nature.

hemisphere: One-half of Earth.

mate: One of two animals that come together to produce babies. Also, to come together to make babies.

spine: One of many pointed parts growing from an animal.

stout: Having a heavy build, especially around the middle.

unique: One of a kind.

FOR MORE INFORMATION

Books

Bodden, Valerie. *Iguanas*. Mankato, MN: Creative Education and Creative Paperbacks, 2023.

Weakland, Mark. *Iguanas*. Mankato, MN: Black Rabbit Books, 2024.

Websites

Iguana
kids.britannica.com/kids/article/iguana/353282
This Britannica website has more facts about the iguana and its life cycle.

Iguana
animals.sandiegozoo.org/animals/iguana
Read more about these lovable lizards from the San Diego Zoo.

INDEX

basking, 10, 11, 15
Caribbean, 8, 20, 21
Central America, 8, 11
cold-blooded, 10
colors, 4, 5, 10, 14, 15, 16
crest, 6, 14
dewlap, 6
eggs, 5, 16, 17, 20
females, 4, 14, 16
Fiji, 8
foods, 12, 15, 18
Galápagos Islands, 8, 13, 18
lizards, 4, 5, 10, 19, 20
males, 4, 14
Mexico, 8
predators, 9, 12, 17, 21
scales, 5, 6, 18
South America, 8
species, 4, 8, 9, 16, 18, 20
spines, 6, 14
tails, 6, 14, 21
tongue, 7
United States, 8